DESTINATION:
JUPITER

DESTINATION:
JUPITER

S E Y M O U R   S I M O N

HarperCollins*Publishers*

PHOTO CREDITS
Permission to use the following photograph is gratefully acknowledged: page 7,
Kyle Cudworth, The Yerkes Observatory. All other photographs courtesy of NASA.

The text type is 18-point Garamond Book.

Destination: Jupiter

Library of Congress Cataloging-in-Publication Data
Simon, Seymour.
    Destination: Jupiter / by Seymour Simon.—Rev. ed.
        p.   cm.
    Originally published: Jupiter. New York: W. Morrow, c1985.
    Summary: Describes the characteristics of the planet Jupiter and its moons, as revealed by photographs sent back by unmanned spaceships.
    ISBN 0-688-15620-7 (trade) — ISBN 0-688-15621-5 (lib.bdg.)
    ISBN 0-06-443759-0 (pbk.)
    I. Jupiter (Planet)—Juvenile literature.   [1. Jupiter (Planet).]   I. Simon, Seymour. Jupiter.   II. Title.
QB661.S585   1998   97-20488
523.45—dc21   CIP
                                                                                                                                AC

12 13 SCP 10 9
❖
For information address HarperCollins Children's Books,
a division of HarperCollins Publishers,
10 East 53rd Street, New York, NY 10022.
Visit us on the World Wide Web! www.harperchildrens.com

To Robert, Nicole,
Joel, and Benjamin

From our planet, Earth, Jupiter looks like a bright star in the night sky. But Jupiter is a planet. It is the fourth-brightest object in the sky, after the Sun, the Moon, and the planet Venus. Jupiter was named after the king of the gods in Roman mythology.

Jupiter is one of nine planets that travel around the Sun. The Sun and its nine circling planets are called the Solar System. Earth is the third-closest planet to the Sun. It takes Earth one year to orbit, or go around, the Sun. Jupiter is the fifth planet, about 480 million miles away from the Sun. Jupiter is so far away that it takes almost twelve Earth years to orbit the Sun once.

Jupiter is the largest planet in the Solar System. It is more than one-and-one-half times as big as the other eight planets put together. If Jupiter were hollow, more than thirteen hundred planet Earths could fit inside it.

No one has ever seen the surface of Jupiter. The planet is covered by clouds hundreds of miles thick. We see only the tops of the clouds: bands of reds, oranges, tans, yellows, and whites.

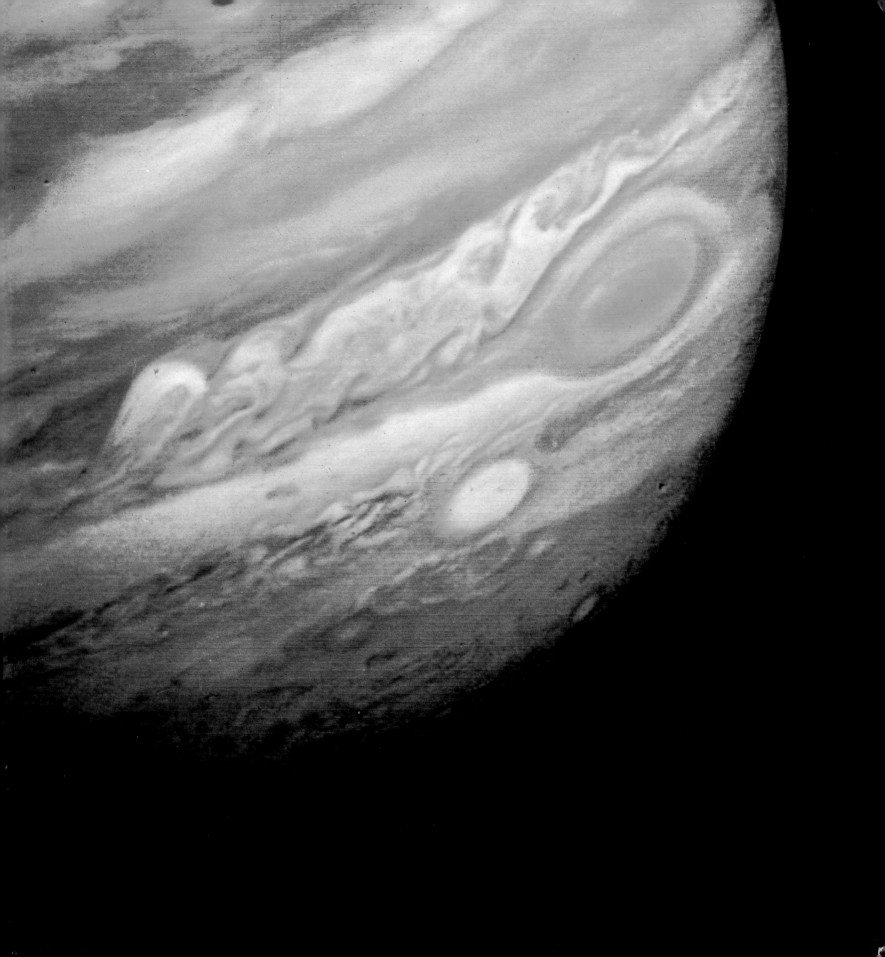

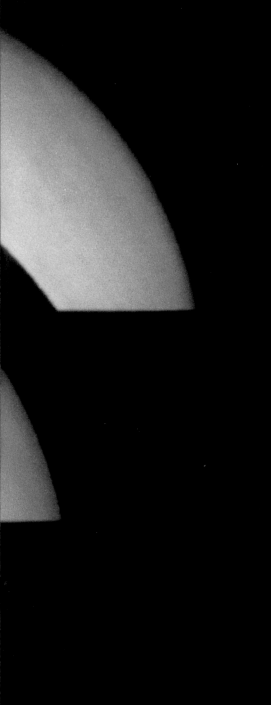

Unlike Earth's clouds, which are made of tiny water droplets, Jupiter's clouds are made of gases, mostly hydrogen and helium. Hydrogen and small amounts of ammonia and methane form the brightly colored cloud tops.

In July 1994, a comet plunged into Jupiter's clouds, with spectacular results. These photos were taken by the Hubble Space Telescope five minutes (lower left) to five days (upper right) after impact. The "bruises" on Jupiter have since disappeared.

The most recent attempt to reach Jupiter was made in 1989 with the launch of the *Galileo* spacecraft. Traveling a hundred thousand miles an hour, the unmanned spaceship took six years to reach Jupiter. *Galileo* had two parts—the main Jupiter orbiter, which was programmed to circle the planet and its moons and study them for at least two years, and a probe that would penetrate the planet's atmosphere.

Galileo arrived at Jupiter on December 7, 1995, and fired its main rockets to place it into orbit around the planet. On the same day, the atmospheric probe plummeted through Jupiter's clouds. This computer-generated photo shows *Galileo* approaching Jupiter and its moon Io.

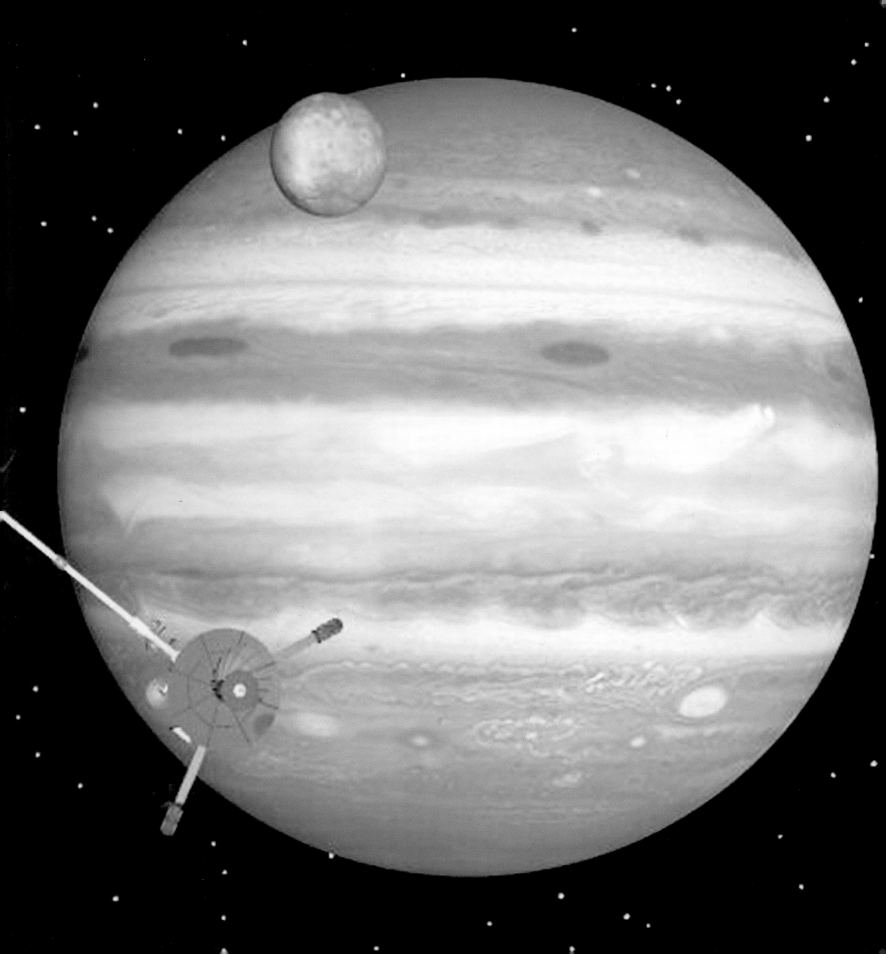

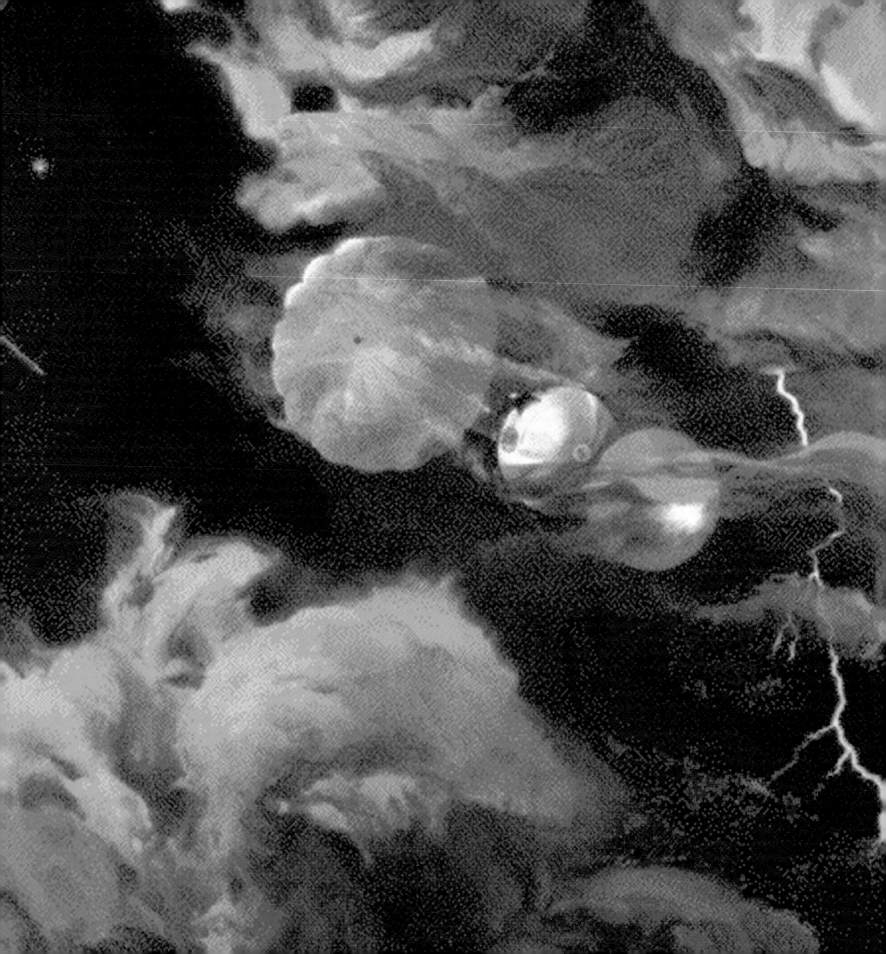

After the *Galileo* probe dived into Jupiter's outermost gases, it opened its parachute and drifted slowly downward about one hundred miles into the atmosphere. For sixty minutes the probe sent back streams of information about temperature, pressure, chemical makeup of the clouds, wind speed, and radiation. Then it lost contact with the orbiter.

The probe revealed that winds blow intensely in Jupiter's top layer of clouds, at speeds of more than three hundred miles per hour. The wind speed increased to over four hundred miles per hour as far down as the probe was able to observe. Scientists think that the winds may blow even more fiercely all the way down at the surface.

Every so often, gigantic bolts of lightning illuminate distant clouds. These flashes are stronger but less frequent than lightning in Earth's atmosphere.

The bright colors of Jupiter's clouds are probably the result of trace elements in Jupiter's atmosphere. Sometimes we can see the lower layers through holes in the upper ones. The light-colored bands are called zones; the dark ones, belts. The colors of the clouds change as you drop down toward the surface. The highest clouds are reds and oranges, followed by yellows, browns, and whites. Blues are the lowest clouds, near the surface of the planet.

Jupiter's surface is an ocean of liquid hydrogen that covers the entire planet. The pressure is millions of times greater than the atmospheric pressure on Earth. Any spaceship landing there would be crushed. At the very center of Jupiter, there is probably a rocky core, about ten to fifteen times the mass of Earth, where the temperature is very hot—about 35,000° Fahrenheit.

If you could stand on the surface of Jupiter, you would weigh more than two-and-one-half times what you weigh on Earth. If you weigh 100 pounds on Earth, you would weigh 264 pounds on Jupiter.

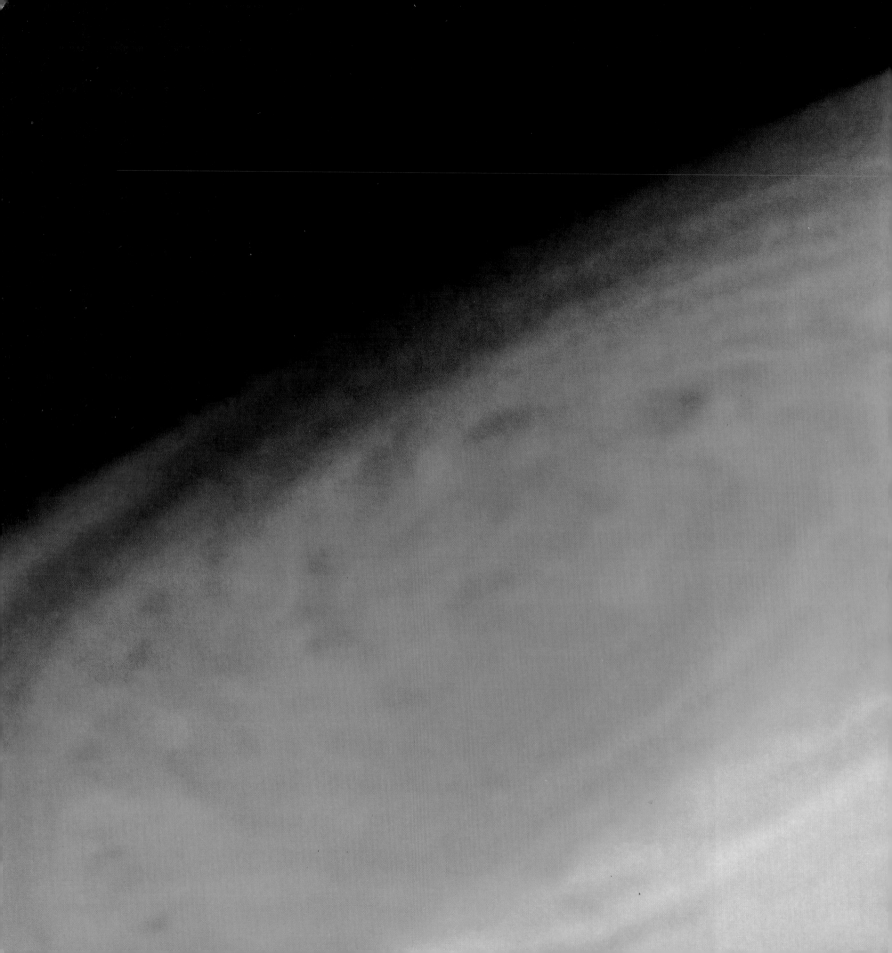

One of the many mysteries on Jupiter is the Great Red Spot. The spot was first seen through a telescope from Earth more than three hundred years ago, but no one knows when it originally formed. It is an enormous storm, a super hurricane big enough to hold two planet Earths inside. This view of the Great Red Spot was taken by the *Galileo* spacecraft in June 1996.

The spot changes in size and color over the years. Sometimes it is small and pink. Other times it grows and turns bright red. But the Great Red Spot does not change its position on Jupiter, and it has kept the same oval shape for centuries. No one knows why.

Jupiter has at least sixteen moons and more may still be discovered. The outer moons are small, most under fifty miles across. The four largest moons circle close to the planet—Io, Europa, Ganymede, and Callisto. These are called the Galilean moons, after their discoverer, the great Italian scientist Galileo. Galileo first saw the moons in 1610 with his small homemade telescope.

Nearly four hundred years later, the spaceships *Voyager,* in 1977, and *Galileo,* in 1995–96, gave us a close-up look at the Galilean moons. The moons are shown so that you can compare their sizes. Io is slightly larger than our own Moon. Ganymede is larger than the planet Mercury. Callisto is the dark-surfaced moon. Europa is the light-surfaced moon.

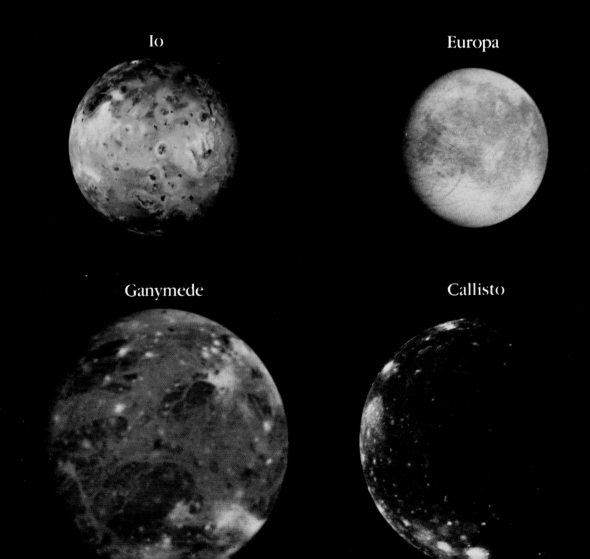

Io

Europa

Ganymede

Callisto

Io (EYE-oh) is the most volcanic moon in the Solar System, with even more volcanic activity than Earth. Io has many different surfaces: volcanic craters, some of which are several miles deep; lakes of molten sulfur; mountains; and lava flows of sulfur or molten rock hundreds of miles long. Some of the hottest spots on Io may reach temperatures of over 1,000° Fahrenheit.

The volcanoes on Io are compared in the images below, from *Voyager,* in 1979 (left), and *Galileo,* in 1996 (right). Prometheus (bright ring in upper right) was first seen as an erupting volcano by *Voyager.* In more recent photos, there appears to be a new dark lava flow from Prometheus as well as other changes in the volcanoes.

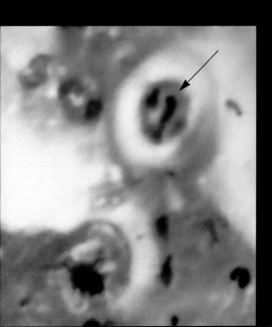 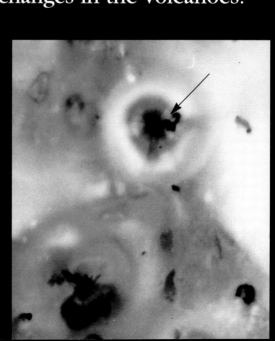

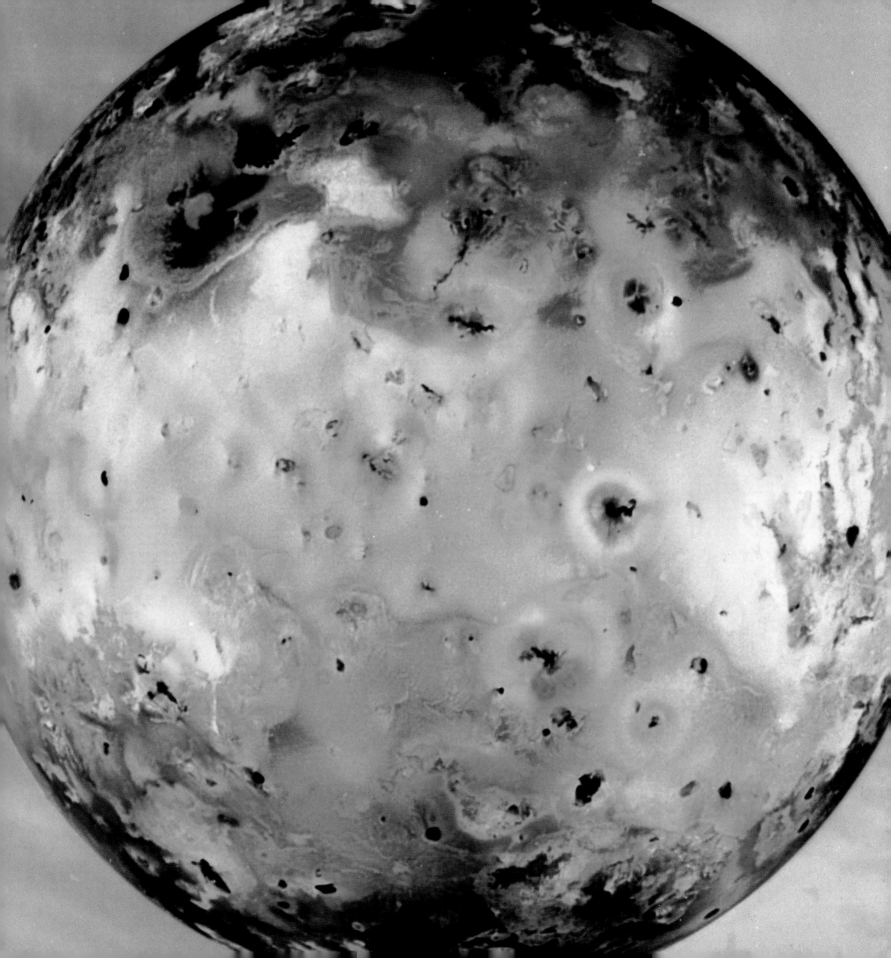

Europa (you-RO-pa) is a bit smaller than Earth's Moon. From a distance, it looks like a smooth white ball. Europa's surface is covered with brown-tinted ice. In some areas, the ice has broken up into large pieces that have shifted away from one another but fit together like a jigsaw puzzle. Scientists think that more ice or liquid water may lie beneath the cracked icy crust.

If an ocean of liquid water does lie under Europa's icy surface, it would be the first place in the Solar System other than Earth where large amounts of liquid water have been found. In fact, Europa has long been considered by scientists to be one of the few places in the Solar System that might be warm and wet enough to support life.

This detailed image taken by *Galileo* shows some of the blue icy plains; white ice floes, ridges, and grooves; and the long dark bands that stretch like highways across Europa's surface. To take this photo, *Galileo* flew within a few hundred miles of Europa.

Ganymede (GAN-a-meed) is the largest moon in the Solar System, two-and-one-half times the size of our Moon. Information and photos from *Galileo* suggest that Ganymede has three layers: a small iron or iron-and-sulfur core, a rocky mantle surrounding the core, and an icy shell on top. Perhaps Ganymede was very hot beneath its surface millions of years ago. The heat melted the ice into water, which then seeped upward, onto the surface, and froze again.

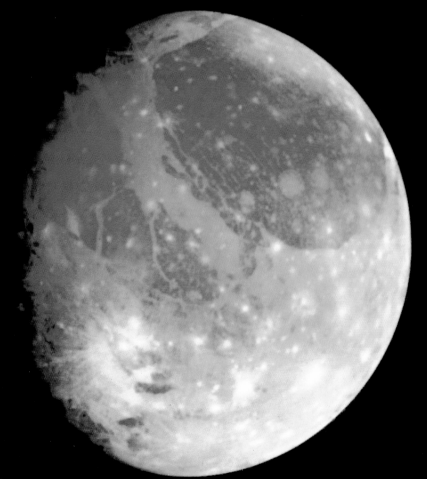

ridges of ice mingled with dust and rock. Dark grooves or valleys lie between the ridges.

This three-dimensional view of Ganymede's surface was made by combining several images taken by *Galileo*. The icy ridges are a few hundred feet high, several miles wide, and hundreds of miles long. The blue-sky horizon was added by the computer; it does not really exist on Ganymede.

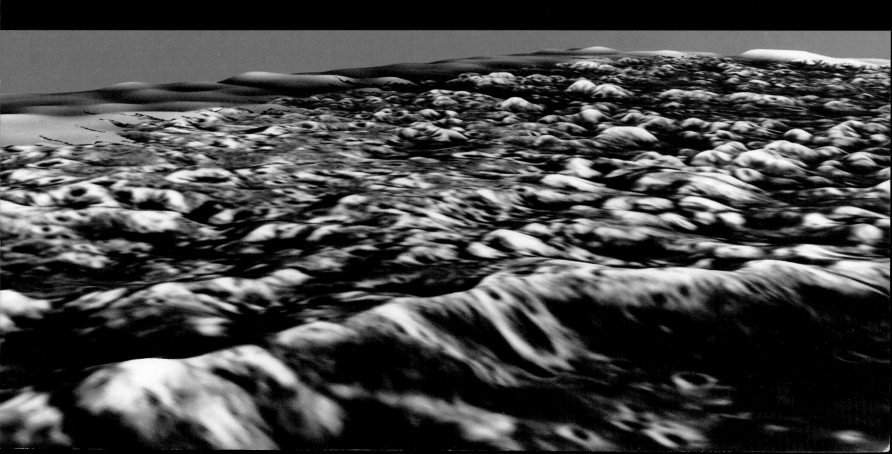

Callisto (ka-LISS-toe) is the outermost of the four Galilean moons. Callisto, like Ganymede, seems to have a rocky core surrounded by ice. But unlike the other large moons of Jupiter, the surface of Callisto is completely covered by craters. It has more of them than any other moon in the Solar System.

Craters are holes in the ground with ring-shaped walls around them. The craters were formed several billion years ago when meteors—giant rocks from space—crashed into Callisto and melted the surface ice. Like rings of water made when you drop a rock in a pond, slushy waves of ice spread for hundreds of miles and then froze in the positions they have today.

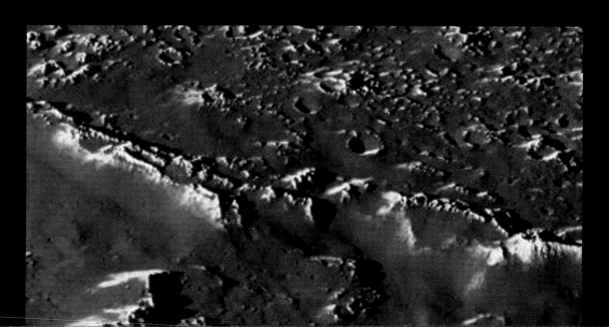

Three images of Callisto were put together to make this photo. The images were sent by three spacecraft: by *Voyager 1* (left side), by *Galileo* (middle), and by *Voyager 2* (right side).

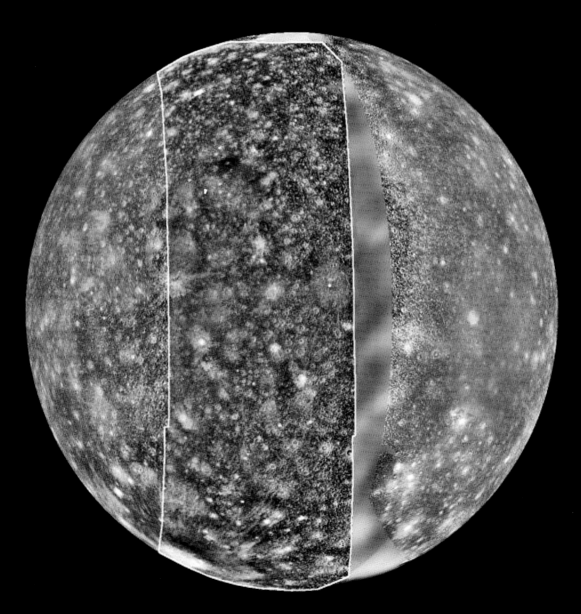

In 1979, *Voyager 1* discovered that a thin ring circles Jupiter. Saturn, Uranus, and Neptune also have similar rings. Jupiter's ring—really several very thin rings that look like a single ring—is much smaller and darker than Saturn's. Particles in the ring don't stay there for very long, because Jupiter's atmosphere and magnetic fields drag them down. Two small satellites that orbit within the ring, Metis and Adrastea, may supply new particles.

This photograph shows Jupiter and part of the ring (right). Sunlight, coming through the ring from behind, is scattered by the tiny particles in the ring. This is like a puff of dust in a darkened room; you can see the tiny particles only if a beam of sunlight slants through the darkness.

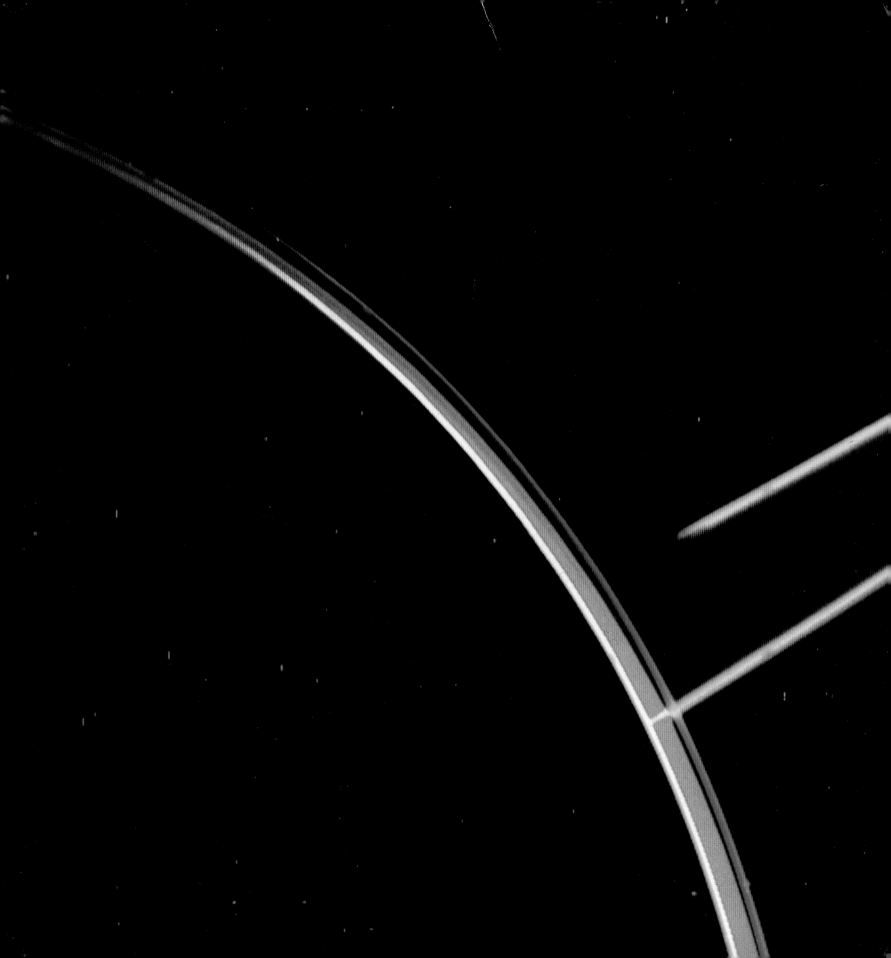

Jupiter and its moons are still an alien wilderness in deep space. It is hard to imagine a place more unlike our own planet Earth. But we are viewing Jupiter and its moons in new ways with *Galileo*. Possibly other spacecraft in the future will make their destination Jupiter. We are at an exciting time in our exploration of our Solar System and the universe.